WHAT IS THUNDER AND LIGHTNING?

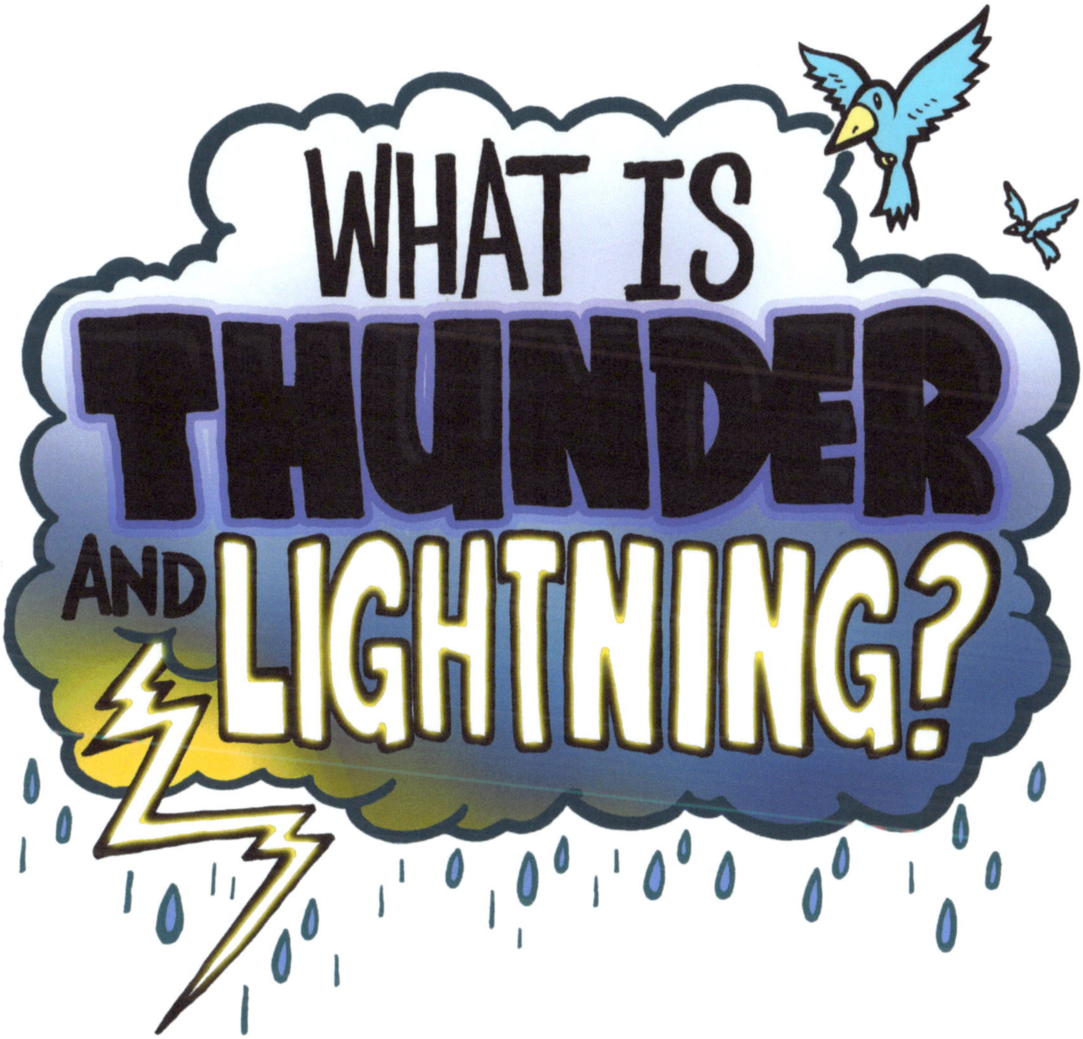

HEATH BRADBERG

Illustrations by Erik Reichenbach

What is Thunder and Lightning?
Copyright © 2022 by Heath Bradberg

First Edition

Paperback ISBN: 979-8-8229-0271-8

For everyone afraid of thunder and storms, I hope this helps with understanding how and why the sound is loud. And for Erik, for helping make this become a reality. This would still be on a scrap piece of paper if you didn't give life to the words. Thank you.

What makes thunder and lightning?
Let's start at the beginning. Water
heats up and rises in the air as
teeny, tiny water droplets, known
as water vapor or steam.

The higher in the sky we go, the colder the air gets, so the water vapor cools down.

Soon, we have little clouds forming.

More water droplets get added to the clouds, and the clouds get bigger and bigger!

When clouds get bigger, they have more energy.

The water droplets at the top of the cloud turn to little pieces of ice, and the ice pieces bump and rub into each other, creating a small electrical charge.

FLASH

When there is a lot of electricity in the cloud, we see a bolt of lightning. But where's the thunder?

BOOM!

CRACK!

RUMBLE!

Why does thunder sound scary?

BOOOM!

FLASH

Thunder is the sound of lightning breaking parts of the air. It's like ripping a piece of paper in half, but much louder.

The closer the lightning, the louder
the thunder.

BA BOOM!
BA BOOOM!

The thunder is a warning for us to head inside and be safe.

We can go back outside to enjoy the puddles, but we have to wait thirty minutes after the last thunder, just to be safe.

Activity to try: We can't see thunder, since it's a sound. For this activity, you will need a piece of paper. Close your eyes and rip the piece of paper right next to your ear. Does that sound loud?

Now take the paper and rip it again, but this time, rip it as far away from your ear as you can reach. Was that more quiet? Thunder works the same way, the closer it is, the louder it will be.

About the Author:

Heath Bradberg is a meteorologist from Stevens Point, Wisconsin. Graduating from St. Cloud State University in Minnesota, Heath went to work at a few tv stations in Wyoming and Texas. Currently, he lives in San Angelo, Texas with his partner and three dogs doing real estate and working on his next book.

About the Illustrator:

Erik Reichenbach is a comic illustrator and graphic designer from Pinckney, Michigan. His artwork has been featured in Entertainment Weekly, People, and San Diego Comic-Con International. He also was a two-time contestant on the CBS reality show Survivor in 2008 and 2013. Currently Erik lives in Ypsilanti, Michigan with his wife, one year old son, and two golden retrievers.